A CIRCLE
OF DILEMMA

The Dramatic Adventures of a Futuristic Scientist
-a science fiction series -vol. 1

Mercy Offor PhD

A CIRCLE OF DILEMMA

Volume 1

The Dramatic Adventures of a Futuristic Scientist

Science Fiction

By

Mercy Offor, PhD

A CIRCLE OF DILEMMA

A Science Fiction Series

Volume 1

Dedication

This Science Fiction Series is dedicated to my son

Emmanuel Offor, Jr.

About The Book

A Circle of Dilemma is a humorous Science Fiction. It is based on the ambitions of a scientist of the future who uses his knowledge of the emerging science - Nanotechnology - to go over and above in the engagement of his students.

A Circle of Dilemma, as the title suggests, is a continuing adventure series that is as thrilling as it is exciting. Once you get on the bandwagon, it is hard to get off as the thrills and laughs keep on coming.

The lead character is a learned professor of science experienced in innovative and cutting edge procedures. The setting of the science fiction is a college campus and students are main characters in the story. This title is educating, engaging and full of lively humor and entertainment.

This is a non-fiction which is recommended for all ages especially for students and lovers of science and anybody with a vivid imagination.

A Circle of Dilemma is the story of an aging scientist's obsession about environmental concerns and issues, especially pollution, climatic changes and clean- up of the atmospheric gasses.

His obsession with emerging technologies, alternative energies and technological products drove him to attempt radical moves designed to dive into scientific discoveries that rightly belong to a future era.

In his zeal to leave his indelible mark on the sands of time, put an end to speculations, and explore more and more innovative ground -breaking research of the future, he sets off a circle of dilemma.

Table of contents

CHAPTER 1

The Scientist of the future and Future scientists at work

The earth quake was not the only thing going on that night. A steady stream of a colorless gas was rushing out of the ground at a number of locations. The carbon dioxide that was stored underground was released by the earthquake. In addition a massive explosion had devastated the city college earlier that night as the end of year activities and dinner came to an abrupt end.

That was the night that Professor San Francis disappeared.

Was the disaster a result of concerted efforts to curb carbon dioxide emission in the city and save the globe! Did the disposal of carbon dioxide underground build up enough pressure to cause an earthquake? This natural or manmade disaster brought havoc to the peaceful city overnight. How did scientific advancement intended to bring enhancement of the way of life of the citizens threaten their destruction? What caused the explosions at the city college?

It is all the ambition of an aging scientist.

Professor San Francis, a retired professor of environmental science lived alone in his medium sized luxury home in the hilly country side at the outskirts of Francisco City. The neighbors thought he was quite wired and mysterious. He did not seem like a social individual and talked to himself most of the time. He had served in highly rated institutions of higher learning and had a number of publications and groundbreaking research under his belt. During his career as a university professor, he had won Nobel Prizes for excellence in teaching and research. However not much was known about his family.

"I need to find a part time job to occupy myself. Moreover, I have a lot of great ideas about curbing carbon dioxide emission and climate change remediation that I must pass on to the next generation," he said to himself, "It will be a pity to die and deny the coming generations of all my ideas about curbing global warming, alternative energies and beyond."

"I will negotiate a part time job in the near-by college, the Francisco City College tomorrow," he said to himself. The city college is a two year college from where the students proceeded to four year colleges and universities. With a background of many years of teaching in big universities, Prof San Francis expected a less demanding and less challenging workload which will match his advancing age. Moreover he will be working only part time, and would accept only one science class. He needs to have enough time for routine exercises and adequate diet to ensure fitness and graceful aging. His acceptance at the city college was a matter of course. The administrators were pleased to offer him the position he was asking for after reviewing his highly impressive resume.

So began the retired professor's part time job that grew larger than him, larger than the entire city!

Being a renowned professor with a number of Nobel prizes, the professor's classes filled up very fast as soon as it was announced. The new course by the retired professor was called *Current Affairs in Environmental Science and Health*, a beginner's course.

"I am really excited about the new course," David, a first year student said to his friend Vicks, another first year student as they strolled to the 500 lecture theater for the opening lecture.

"Taking any course at all under Professor San Francis who is an icon is a good idea," replied Vicks, "I will pay attention and make sure I understand every word and comply."

The two friends had no idea what was in store for them and the rest of their college mates!

The first set of lectures given by the Professor centered on curbing carbon dioxide emission.

"Carbon dioxide constitutes one fifth of the atmospheric gases. It initiates the process of photosynthesis which is the start of the food chain in the ecosystem," he told the students.

"Why do you want to curb the emission of a gas that is so important?" asked a student.

"It is also one of the gasses that reflect back much of the heat escaping from the earth's surface thus making the globe warmer" replied Prof San Francis, "it is one of the so called greenhouse gasses."

"As you will recall, the most active, but less abundant greenhouse gas is Methane, the simplest hydrocarbon, which is given off from marshes, organic materials associated with wetlands and livestock manure. The most important greenhouse gas which is also a component of atmospheric gasses is Carbon dioxide; released through natural processes such as respiration and volcanic eruptions and through human activities such as deforestation, and burning fossil fuels. Others include Nitrous oxide, water vapor and Chlorofluorocarbons -CFCs - from Aerosol Sprays and Refrigerants" the professor explained.

The concept of greenhouse gasses should be prior knowledge for the students, but the professor took time to refresh their memory on a topic that he considers crucial as prerequisite to the topic of alternative energies that he is going to take up next.

"Greenhouses –with glass walls and glass roofs - also called glasshouses or hothouses – are glass buildings in which plants are grown so that they are protected from cold weather" he further explained, "because the glass allows the sunlight to go through unhindered, but traps the short wave infrared light. These waves that cannot escape back into space through the glass turn into thermal energy which warms the greenhouse. Especially in cold climates, greenhouses provide warm nurturing environment for growing plants, tomatoes etc."

"Just like the greenhouses, methane, carbon dioxide, nitrous oxide and other greenhouse gasses reflect back solar radiation escaping from the earth. Most climate scientists agree that the main cause of the current global warming trend is human expansion of the *greenhouse effect*— warming that result when the atmosphere traps heat radiating from earth toward space. This expansion is due mainly to release of carbon dioxide through deforestation and burning fossil fuels. Humans have increased atmospheric carbon dioxide concentration by a third since the Industrial Revolution began," the professor expanded.

"This is similar to what happens when we park our cars under the sun on a hot summer day and it gets too hot to get back in" added a student. "Exactly," agreed the professor "the glass windows in your car acts like the glass walls and roof in a greenhouse;–allowing the sunlight to enter the car but hindering the heat component of the light from passing through the glass back into the atmosphere."

"Innovative breaks are surfacing all over the globe," the professor continued. "European sources talk about a cutting edge technology in the making: *The Carbon dioxide Compression and Storage Project,* to address this issue of global warming. Alternative energies are springing up all over the world to decrease the amount of carbon dioxide in the atmosphere. On individual basis, a variety of measures are becoming a life style. Evidences of global warming include changing weather patterns, severe weather world-wide such as tsunamis, flooding, and incessant hurricanes and tornados! These red flags include the melting of glaciers at the North Pole which threatens to bring two big killers together – the grizzly bear and the polar bear!"

"In the atmosphere, climate change makes the presence of extra carbon dioxide undesirable, elsewhere, this gas is chemically active. It is an acid forming gas which is soluble in water, dissolving to give carbonic acid. For the aquatic ecosystems, this is not good news. Increase in the acidity of the oceanic or sea water will endanger the fish families, shrimp families and countless others. Disposal in the waters – lakes, rivers and oceans - is therefore not a viable option for carbon dioxide."

"What about disposal of carbon dioxide underground?" The professor went on to say. "This sounds like a one two punch. First, acidic soil would devastate the roots of plants and trees and lead to the phasing out of our green plants. The green plants and trees remove carbon dioxide from the atmosphere and decreasing their population will consequently increase the atmospheric carbon dioxide. In addition, compressed gas being forced into the earth might set off earth quakes."

The professor said that it would take a long time, possibly several decades to build up sufficient pressure to cause problems such as earth quakes. There is no immediate threat and presently the risk is minimal.

He went ahead to advise the Environmental Watch Guards - a group that monitor and regulate emissions - to dispose emitted carbon dioxide underground as a temporary measure. Soon, the disposal of carbon dioxide from coal and petroleum burning industries were channeled underground in Francisco City using large pipes.

"Our next series of lectures would center on the alternative sources of energy that are being researched as replacement for the fossil fuels - *crude petroleum, coal and natural gas*," Prof San Francis announced, as the lectures in the first week came to an end, "we shall look into the numerous cutting edge research springing up in our next series of lectures when the class meets again next week."

The excited students left the class filled with anticipation. They were very happy about being in the group that is making a difference. Already the recommendation of the professor regarding the disposal of carbon dioxide gas, received very favorable review and was being followed by the city governing body. Carbon dioxide gas was being disposed underground and no longer in large bodies of water to scale back global warming.

During the next series of lectures, in the coming weeks, the students learned about Solar cells being used to harness solar energy and Fuel cells that generate electricity. The professor dwelt passionately on the use of extremely small particles of metals known as *Nano Particles* to reduce the cost of expensive precious metals being used in the Cells.

"Did you notice the passion of Prof San Francis for the Nano particles"? Vicks asked David after the first three classes on alternative energies.

"Yes" David replied, "I wonder exactly what they are and why he seems very passionate about them."

I guess *we shall understand it better bye and bye* as the song writer wrote. These series of lectures are shaping up to be both mysterious and enlightening.

Chapter 2

Nano Enhanced Artificial Intelligence

If the students were noticing something odd about the professor already, it is too early to comment and he was just getting started. He explained that the use of Nano particles of precious metals speeded up the rate of the reaction, when used in the membrane of the fuel cell. This is because the very tiny particles of size *ten to power minus nine* (one billionth) presented larger surface areas for the reaction. As he explained he seems to be getting more and more excited about the faster speed of catalysis afforded by the Nano particles of the precious metals used.

"I feel really elated about this area of research because increased speed is desirable to get answers and get ahead in all areas of life," he explained.

"It appears that the word *delay* has been eliminated from his dictionary," another eager student commented under his breath.

During the next few weeks, Prof Francis speculated on the idea that the enhancement of the fuel cells by nano-sized particles can be applied to other areas and activities of man. "Now in the course of my professorship in the university, I did enormous research in the area of precious metals and the mechanisms of their reactions especially catalysis. I expanded my research into the formulation of *novel* precious metals that have more than twice the speed of the known precious metals when it came to catalyzing a process. These set of novel precious metals appear to open up brand new mechanistic pathways," concluded the professor.

"Which precious metals are you talking about," asked a perplexed student, "I am somehow lost in the midst of these *novel precious metals,* do they have a name?"

"Oh yes," replied the professor, smiling from ear to ear. "I synthesized the novel elements while working alone in my laboratory - *my babies* indeed they are- and so I named them after myself. They are known as the *San Frances group of precious metals*. They are brand new artificial precious metals. Their atomic numbers have not yet been determined, and consequently their would-be positions on the Periodic Table of Elements are yet to be determined."

"So we shall not be able to locate these novel precious elements on the periodic table," said a student, making the comment in a low voice. "I considered my knowledge of the Periodic Table of Elements quite decent up till now, but the professor sure put a dent on it with this twist" said another student thinking aloud.

The professor continued to talk excitedly about the Nano particles from his newly discovered precious metals. He seemed to suggest that these novel Nano particles will influence/catalyze the human thought process the same way that known Nano particles catalyzed the reactions that generate electrons in *the cell permeable membrane* of fuel cells. If the novel particles from his newly synthesized precious metals found their way into the human brain, it would certainly improve alertness and intelligence.

Right away he set out to try the implantation of small chips of his novel precious metals in rats. He used the chips and Nano particles from three out of his five San Francis

precious metals. The rats with the implants were kept in a special cage side by side with rats without implants that acted as control!

"How are the rats implanted with the Nano chips doing, Professor?" the students will ask at the start of each class.

"I have decided to bring the *guinea pigs* and the control specimen to the laboratory. This way, you can all compare and document the results of this fascinating experiment as part of your practical work and science project in this course," the professor replied.

At least a dozen rats were in each cage which was promptly transported into the environmental science laboratory. The students went in on a daily basis to check on the progress of the two categories of *specimens* as the rats were referred to. There was indeed no comparison between the two groups of rats. The rats with the implanted chips were faster, more aggressive and were developing much faster and becoming a lot larger. This is exactly what the professor had hoped for. In addition to increased and sharpened intelligence, the rats with the nano implants doubled their size within a few weeks and continued to grow. Increasing excitement mounted over the new experiments.

The professor was ready to take the next step. The next step he envisioned would be the implantation of the nano chips in humans, especially in people of below average height and below average intelligence he speculated.

"I would recommend the implanting of nano chips in humans before the end of the next semester," he told the environmental science students in his class.

"Professor San Francis himself is not that tall," observed a student when the class session was over and the students laughed as they headed for lunch in the cafeteria.

"That may be his motivation for the investigation into the growth causing chips," another student pointed out.

Next class the professor added another set of experiments using a species of raccoons.

This time he prepared nano chips made from an alloy of two of his San Francis precious metals to observe the effect. As before, he had a second set of raccoons as control. He assigned groups of students to observe the implanted and the control set and to feed them with various nuts thrice daily while recording their temperatures hourly.

"The raccoons I have implanted with metal alloy are to be closely monitored," he emphasized.

"What is an alloy?" asked a curious student. "Oh, an alloy is a mixture of two solids. The definition for an alloy is a solid-solid mixture," explained the professor.

The observations were started promptly and the students took turns to go into the laboratory to collect the required data.

"There are no raccoons in the cage for the implanted raccoons," a student reported the following morning when he arrived to feed the animals.

The students searched around to see if the experimental raccoons were moving around the laboratory or inside the lockers; perhaps a student forgot to lock the cage after feeding them and they came out, but they were nowhere to be found.

"How could the animals escape from a securely locked cage? There simply has to be another explanation," the Professor said when he was called to the experimental room. "I would propose another hypothesis to explain this observation."

"What is your hypothesis, Professor," asked one of the students.

"I propose that the raccoons are still inside the cage but simply invisible to the naked eye due to the effect of the nano alloy chips," replied the professor.

"With all due respects, Prof, you are not making any sense," the students replied.

"What I mean to say is that the raccoons disappeared due to the effect of the Nano alloy chips. Hum, I wonder when they will reappear."

As the professor spoke, the raccoons reappeared one by one inside the cage!

"How cool is that!" shouted the excited students. "More, we want more!"

"Can you get more of the alloy chips implanted in them to repeat this disappearance drama?" they asked.

"I think it will be even more exciting to try the chips a little higher in the food chain," replied the professor. I am thinking of trying dogs and cats.

A frantic search for dogs and cats began almost immediately. Not many wild dogs and cats are wondering about in the neighborhood or bushes, so the professor dashed to the animal shelter to gather enough *specimens*.

Super – intelligent cats and dogs

A new breed of super intelligent dogs and cats emerged as a result of Prof San Francis experiments, as well as disappearing dogs and cats.

The first set of 30 cats and dogs were put in three large cages. The first cage contained five cats and five dogs with no embedded chips. The next cage contained the five dogs and cats embedded with nano chips and the third cage housed the five cats and dogs with the nano alloy.

Large quantities of tinned dog and cat food were loaded in one corner of the laboratory and gallons of water and milk were also made available for the animals. Excitement mounted as the students took turns to feed the animals three times a day and carry out their observations.

The adventurous professor monitored the observations very closely and left directions that the first sign of disappearance of the set of the animals in the third cage embedded with the nano alloy chips should be reported to him.

Day after day, the students watched and wrote down their observations in their note books throughout the first week of the observations. At the start of the second week, the third set of the animals disappeared and did not appear till the following day. Professor San Francis could not be happier.

"This is just what my calculations predicted," he exclaimed, "that gives me enough time to get ready" he added under his breath.

"Enough time to do what?" David, one of the students who watched the professor very closely asked. He got no reply. The Prof. only looked at him and smiled.

"I will add one more dimension to this intricate experiment to provide the last piece of this puzzle," he announced.

The students were not sure what he meant by *this puzzle*.

"Tomorrow, I will bring in two pairs of dogs and cats for cages two and three, both conditioned with the nano powder and nano alloy powder in their drink, respectively. Observation of the new specimens will begin immediately and every new manifestation will be reported right away. This is very important," he announced.

All the new cats and dogs that have been given the oral dose of the nano powder and nano alloy powder in their drink were painted with blue paint all over their body including their faces for identification.

The two environmental science students who were the most ardent about these novel and intriguing set of experiments by Prof Francis, David and Vicks, discussed this latest trend in the experimental procedures.

"Why do you think the professor is investigating the addition of the powerful chips in powdered form in their drinks and calling it the last piece of the puzzle"? David asked.

"I am not ruling out the possibility that the professor would like to try his innovative experiments on human beings," Vicks replied, "let's wait and see as this intricate drama keeps unfolding."

In the course of the next week, the blue painted animals who were nicknamed *The Blues* by the students showed the same manifestations as the animals that had the chips embedded and disappearance began in exactly one week. They only took a little longer to reappear, reappearing in five days instead of one day.

"Perfect!" "This calls for a celebration," the professor exclaimed jumping up and down with excitement and exhilaration. Some of the students joined him in a wild dance of celebration. He put some of the bottles of the nano metal powder and nano alloy powders in the cupboard inside the chemical store room.

"There is so much more to be investigated. For example, where did the animals go when they disappeared and what capabilities did they have in that invisible state? Are they able to observe what we are doing here and can they influence what is going on with us? Are they above or below us or at the same level as we are? Are they a threat of any magnitude to us or to each other in their invisible state? Can they see each other in that state? As George Bernard Shaw said 'Science never solves a problem without creating ten more' and I can tell you, as a learned scientist, that the more knowledge you acquire, the more you discover what you do not know! Many more questions are raised by this disappearance, but we shall officially stop the investigations at this point," he concluded.

Chapter 3

End of year dinner party

The next week was the final dinner party for faculty and students to mark the end of the academic year and give recognition to students who have excelled in their various disciplines. The professor had earlier notified David and Vicks that they were nominated for the environmental science awards so they must be sure to attend the ceremony.

All invitees took their seats at the ceremony and socialized as refreshments were being served and more students and faculty arrived in the conference hall. David and Vicks arrived early and eagerly awaited the arrival of Professor San Francis, but he did not arrive. They waited anxiously and wondered why. Should they go to the science complex and find out if he was at work and why he did not answer his phone? Fearing they might miss the awards ceremony if they left, they decided to join the other students and staff and pick up their awards.

The President welcomed all present and called the award winning students who picked up their awards faculty by faculty.

Soon all were seated at the well decorated tables for the dinner party. The atmosphere was that of a very merry celebration and everyone was excited and full of smiles.

Suddenly the chatting and laughing and excitement were interrupted by a figure walking in with giant strides more than double the size of the tallest person in the gathering.

"Who or what is that?" "Oh my goodness," "This is not happening," people exclaimed wiping their eyes to ensure they are not seeing double.

The figure walking in was as ridiculous and as comic as ever. The legs of the super tall gentleman were longer than a normal human being. With no trousers of that size available, the normal trousers he had on looked like mere underpants. The shirts meant for a normal person looked like a grossly undersized singlet.

The activities grinded to a halt and all eyes were fixed at this figure as it strode in apparently unaware of the impression his appearance made on the assembly.

"Can this possibly be Professor San Francis? "Did the professor have some of the nano powder in his tea without our knowledge?" David asked Vicks who stared at the figure with wide frightened eyes.

The figure, looking even more comic with a weird smile on his face, came straight towards David and Vicks. It was indeed Professor San Francis. He tried to sit down with them at their table, but that was not possible. The nearest description to what he looked like is a clown walking on stalks in a circus. The first attempt he made at sitting down ended with the table turned over kicking everyone in the direction of his legs down from their seats.

The attendants quickly devised a plan. They kept a seat at the far end of the conference room and directed him to that seat. To reach what was served at the various tables all he had to do was stretch his hands which went across two or three tables nearest to his end of the room.

The arrival of Professor San Francis in this kind of grand style added a whole new dimension to the end of year celebrations and to innovative and groundbreaking research in science. The president quickly called for an end to the celebrations as soon as the dinner was over and called an emergency meeting of the staff right away in the same venue. All the students quickly left the scene and the staff gathered round one of the tables at the far end of the conference room from where Professor San Francis sat.

Though he was not within a normal earshot of the discussing group, the professor with enhanced auditory systems due to the nano powders taken in his tea heard all their discussion. The discussion centered on how to get rid of him from the faculty with immediate effect!

This was not at all palatable to the professor who cannot see what he has done wrong. He has not broken any law to the best of his knowledge. He therefore does not deserve a dismissal or any of the stuff they were deciding to do to him. Are they discriminating against him now? This certainly made him mad.

"You all are certainly going to get it," he thought as he noiselessly got up from his seat. He strode towards the meeting members and stretched his hand towards them. He grabbed one person in each hand and threw them forcefully against the wall knocking them out instantly. Next he grabbed another two and threw them mercilessly against the walls. Some were wounded, some fractured. He made for the next two as a pandemonium erupted and all ran for their lives from this superhero which Professor San Francis has become.

The super long hands continued relentlessly grabbing two faculty members at a time and dashing them silly or fractured against the walls.

One person who was lucky to make it out of the conference room called frantically to the police "He will kill us all –the giant – in the college conference hall – HURRY!"

Minutes later, the sirens were heard all around the campus as the squad cars raced to surround the conference hall. Rushing into the conference room with their weapons drawn, the police searched in vain for the *giant killer*.

With his heightened hearing and intelligent systems, Professor San Francis heard the phone call to the police and promptly disappeared.

The severely wounded and fractured staff members were taken out with stretchers to the ambulances that accompanied the squad cars.

Meanwhile the invisible professor watched all the drama making sure no one came too near him.

His two most ardent students, David and Vicks who lingered about near the conference center while other students went home were horrified by all that took place. When they did not see the supersized professor coming out in hand cuffs with the police, they had a good idea what had transpired. They did not want their mentor and hero arrested or stopped by anyone. They both smiled to themselves and thought "He must have taken both the nano powder in his tea as well as the nano alloy powder and could disappear at will in addition to being supersized."

The aspect of the entire episode that resonated too well with them was the idea of escaping the police or any other authority by disappearing. Don't we all wish we could disappear when things turn ugly and embarrassing?

"Cool" thought the two boys and proceeded to the laboratory storage room to pick up the bottles of the two types of nano powders that the professor had stored in there. They quickly pocketed all the bottles of the novel San Francis precious metals and hurriedly left the laboratory and headed home.

The two friends got away just in time.

Professor San Francis now having the exclusive advantage of being invisible came out of the conference room. He took a good look at all the confusion and all the fractured and wounded being put into the ambulance and then headed to the laboratories, unseen by anyone since he is invisible. The meeting about his dismissal really made him furious, especially as he does not have a family and no other means of livelihood. He was not done yet with his revenge. He wanted to register his disapproval of the decision being taken about him by the faculty in a way that they will not soon forget!

He went into the laboratory and straight to the chemical store room and picked a very large chunk of sodium metal. He then proceeded to the large tank reservoir filled with water. He only had to stretch his long hand to reach the top of the reservoir and drop the chunk into the large tank.

A deafening explosion rocked the entire science complex, the conference area and surroundings, demolishing a section of the science complex!

The professor was nowhere to be seen!

He simply vanished, leaving many unanswered questions. "Was he killed in the bomb blast he initiated or did he manage to escape? If he escaped, was he unable to get back to normal size and to visibility? The human body can remove a precious metal from the body through the excretory organs over time if it does not undergo radioactive decay. Is this taking an indefinite length of time to occur for Prof San Francis?

Chapter 4

A Band of Friends

The rest of that historic night ended with the happenings at the college campus as the major headlines in the news casts and evening papers. The earthquake that rumbled from the city limits to a quarter of the area at the west side of the city had ended, and the drama and disappearance of the university professor was the breaking news. The city had known about natural disasters and indeed frightening weather conditions that pointed to global warming, but the disappearance of humans, or indeed of any other species, was rather disorienting news. Only the university staff and students who were present at the end of year party had seen the giant - sized figure with the features of Professor San Francis which eventually disappeared. The first responders, the 911 police unit, saw nothing and had only the tale of the giant by the professors and students to rely on. Were these university staff and students seeing visions?

Is this a fact or fiction? The police were not quite sure.

The two students, David and Vicks who were his faithful followers are not giving up. They were determined to go on searching for answers using themselves and maybe some other unsuspecting humans as specimens or maybe until they also disappeared as well. The questions that the professor had raised about exactly what the dogs and cats did when they disappeared and possibly where they went must be answered!

The two students, now friends and confidential partners, met briefly to map out a plan of action that night. Despite the mounting confusion and nightmare, both decided to get together a band of friends – five more students –to make a gang of seven to join their investigation. None of their parents would be told of this investigation and all their resolve in the secret meetings or the laboratory work they plan to carry out in the days, weeks, months or even years to come. Both were saddled with very important assignments; to think about which students should be approached to join their gang and an appropriate venue for their meetings and possibly dangerous experimentations.

The two friends did not have many problems identifying the five students most suited to join their team. Steven never paid much attention in class and is a *do- exactly- as -I am- told* kind of guy. He will not be difficult to control. He has scarcely scored over 40% in any science class, except when he copied right off the next student's answers. Nathan is a below average height or outright *short –older- guy* and if there is any chance that the said nano chips or particles will make him any taller, you can count him in!

Raphael is very friendly and very smart. He enjoyed the series of lectures on nanotechnology and hopes to pursue a science based career in university teaching or computer engineering. The only fly in the ointment as far as he is concerned, is his girlfriend Rhoda who seem to be around him all the time, getting him to do all her science assignments and enjoys only English and literature, her favorite subjects. Raphael will be quite a resource person and may reject the offer, if Rhoda is not invited. Unsure whether a woman can keep a secret of the magnitude they are looking to cook up, David and Nick decided to invite her as well. Will she burst the bubble for them by exposing their secret location and dangerous/ creepy experiments? They were not too sure, but were willing to take the chance.

Last, but not the least, they decided to call in Johnny, a rather reserved student whose parents had recently relocated from a distant city in the Midlands. He is an *okay* student but does not have many friends yet. He is likely to jump on the proposition for the sake of having friends.

The choice of a *band of friends* complete, David and Nick really excited, were ready to go bed. They would contact the chosen few the next day and as for the choice of a venue for meetings and lab; "Thank goodness the earthquake opened up a number of *caves* and tunnels," said David who sees himself as the un-named leader of the up-coming team, and Nick had a good laugh. "Well we might just settle for houses or homes on sale," he replied.

The two friends had very high hopes of catching up with and indeed surpassing Professor San Francis in his experimentations on nanotechnology and disappearing techniques; and have dreams of completely amazing the world at large! Will they or will they not?

Thus the circle of dilemma continues!

(Be sure to pick up volume -2 of A Circle of Dilemma, and join the author in the next series of adventures and exciting happenings in this science fiction series)

Sci-Fi Quiz;

1) *Why are some gasses known as the 'greenhouse gasses'? List three of them (pg.10-12)* --

----

----

2) *What is the greenhouse effect?(pg. 12)*----------------------------------

----

3) *What is the definition of a Nano?*

   (pg. 15)   ---

4) *Are the San Francis precious metals found on the Periodic Table of Elements? Are they fictional or real?*

   (pg.16)   --

5) *What is an alloy?*

(pg. 18)   --

ABOUT THE AUTHOR

Dr. Mercy Offor has a doctor of philosophy degree in Organic Chemistry. She began her graduate studies at the University of Ibadan, Nigeria as a university scholar. She carried out research towards her doctorate degree at Clemson University, South Carolina and the University of Alabama In Huntsville, USA. She did her Post-Doctoral research at the University of Warwick, Coventry, England as the recipient of the prestigious Commonwealth fellowship award in 1983.

She began her career as an educator at the University of Benin, in Nigeria where she was a senior lecturer. She obtained her Post Graduate Diploma in Education from National - Louis University Chicago, USA (2002 – 2004).

She now teaches at the high school level in Chicago where she has contributed immensely to the development of the high school science curriculum. In the summer of 2006 she carried out summer research in the Research Experience for Teachers at the University of Illinois, Chicago, working on nanotechnology. In summer of 2007 and 2008, she worked at the Northwestern University on climate change issues. She co-authored a science text for high school science students titled *Dye – Sensitized Solar Cell Module*, addressing the concepts of nanotechnology and alternative energy for high school level students. She engaged in pilot teaching for the Climate Change Education Project curriculum for Middle and High School students, supported by CPS, NASA and Chicago Botanical garden, 2011 – 2012. She was nominated for the Disney award which honors excellence and creativity in teaching, in 2006.

Dr. Offor is the author of the fascinating memoir - *Refreshing Springs, Great Legacies from a Renowned Dynasty* - as well as a number of entertaining novels and science fiction; all available on Amazon.com.